Help Me Understand Genetics
Genetics and Human Traits

Reprinted from https://ghr.nlm.nih.gov/

Lister Hill National Center for Biomedical Communications
U.S. National Library of Medicine
National Institutes of Health
Department of Health & Human Services

Published November 22, 2016

Genetics and Human Traits

Table of Contents

Are fingerprints determined by genetics?

Each person's fingerprints are unique, which is why they have long been used as a way to identify individuals. Surprisingly little is known about the factors that influence a person's fingerprint patterns. Like many other complex (https://ghr.nlm.nih.gov/primer/mutationsanddisorders/complexdisorders) traits, studies suggest that both genetic and environmental factors play a role.

A person's fingerprints are based on the patterns of skin ridges (called dermatoglyphs) on the pads of the fingers. These ridges are also present on the toes, the palms of the hands, and the soles of the feet. Although the basic whorl, arch, and loop patterns may be similar, the details of the patterns are specific to each individual.

Dermatoglyphs develop before birth and remain the same throughout life. The ridges begin to develop during the third month of fetal development, and they are fully formed by the sixth month. The function of these ridges is not entirely clear, but they likely increase sensitivity to touch.

The basic size, shape, and spacing of dermatoglyphs appear to be influenced by genetic factors. Studies suggest that multiple genes are involved, so the inheritance pattern is not straightforward. Genes that control the development of the various layers of skin, as well as the muscles, fat, and blood vessels underneath the skin, may all play a role in determining the pattern of ridges. The finer details of the patterns of skin ridges are influenced by other factors during fetal development, including the environment inside the womb. These developmental factors cause each person's dermatoglyphs to be different from everyone else's. Even identical twins, who have the same DNA, have different fingerprints.

Few genes involved in dermatoglyph formation have been identified. Rare diseases characterized by abnormal or absent dermatoglyphs provide some clues as to their genetic basis. For example, a condition known as adermatoglyphia is characterized by an absence of dermatoglyphs, sometimes with other abnormalities of the skin. Adermatoglyphia is caused by mutations in a gene called *SMARCAD1*. Although this gene is clearly important for the formation of dermatoglyphs, its role in their development is unclear.

Scientific journal articles for further reading

Burger B, Fuchs D, Sprecher E, Itin P. The immigration delay disease: adermatoglyphia-inherited absence of epidermal ridges. J Am Acad Dermatol. 2011 May;64(5):974-80. doi: 10.1016/j.jaad.2009.11.013. Epub 2010 Jul 8.PubMed: 20619487 (http://www.ncbi.nlm.nih.gov/pubmed/20619487).

Nousbeck J, Burger B, Fuchs-Telem D, Pavlovsky M, Fenig S, Sarig O, Itin P, Sprecher E. A mutation in a skin-specific isoform of *SMARCAD1* causes autosomal-dominant adermatoglyphia. Am J Hum Genet. 2011 Aug 12;89(2):302-7. doi: 10.1016/j.ajhg.2011.07.004. Epub 2011 Aug 4. PubMed: 21820097 (http://www.ncbi.nlm.nih.gov/pubmed/21820097). Free full-text available from PubMed Central: PMC3155166 (http://www.ncbi.nlm.nih.gov/pmc/articles/PMC3155166/).

Warman PH, Ennos A.R. Fingerprints are unlikely to increase the friction of primate fingerpads. J Exp Biol. 2009 Jul;212(Pt 13):2016-22. doi: 10.1242/jeb.028977. PubMed: 19525427 (http://www.ncbi.nlm.nih.gov/pubmed/19525427).

To find out more about the influence of genetics on the formation of fingerprints:

The UCSB Science Line from the University of California, Santa Barbara provides information about how fingerprints are formed (http://scienceline.ucsb.edu/getkey.php?key=2650).

The Mad Sci Network offers many Q&As related to fingerprints (http://www.madsci.org/FAQs/body/fingerprints.html), including the genetics and development of dermatoglyphs. The questions were asked by students and answered by scientists.

The Washington State Twin Registry has an FAQ about the fingerprints of identical twins (http://wstwinregistry.org/2015/10/01/do-identical-twins-have-identical-fingerprints/).

OMIM.org provides more detailed genetic information about dermatoglyphs (http://omim.org/entry/125590) and adermatoglyphia (http://omim.org/entry/136000).

Is eye color determined by genetics?

A person's eye color results from pigmentation of a structure called the iris, which surrounds the small black hole in the center of the eye (the pupil) and helps control how much light can enter the eye. The color of the iris ranges on a continuum from very light blue to dark brown. Most of the time eye color is categorized as blue, green/hazel, or brown. Brown is the most frequent eye color worldwide. Lighter eye colors, such as blue and green, are found almost exclusively among people of European ancestry.

Eye color is determined by variations in a person's genes. Most of the genes associated with eye color are involved in the production, transport, or storage of a pigment called melanin. Eye color is directly related to the amount and quality of melanin in the front layers of the iris. People with brown eyes have a large amount of melanin in the iris, while people with blue eyes have much less of this pigment.

A particular region on chromosome 15 plays a major role in eye color. Within this region, there are two genes located very close together: *OCA2* and *HERC2*. The protein produced from the *OCA2* gene, known as the P protein, is involved in the maturation of melanosomes, which are cellular structures that produce and store melanin. The P protein therefore plays a crucial role in the amount and quality of melanin that is present in the iris. Several common variations (polymorphisms) in the *OCA2* gene reduce the amount of functional P protein that is produced. Less P protein means that less melanin is present in the iris, leading to blue eyes instead of brown in people with a polymorphism in this gene.

A region of the nearby *HERC2* gene known as intron 86 contains a segment of DNA that controls the activity (expression) of the *OCA2* gene, turning it on or off as needed. At least one polymorphism in this area of the *HERC2* gene has been shown to reduce the expression of *OCA2*, which leads to less melanin in the iris and lighter-colored eyes.

Several other genes play smaller roles in determining eye color. Some of these genes are also involved in skin and hair coloring. Genes with reported roles in eye color include *ASIP*, *IRF4*, *SLC24A4*, *SLC24A5*, *SLC45A2*, *TPCN2*, *TYR*, and *TYRP1*. The effects of these genes likely combine with those of *OCA2* and *HERC2* to produce a continuum of eye colors in different people.

Researchers used to think that eye color was determined by a single gene and followed a simple inheritance pattern in which brown eyes were dominant to blue eyes. Under this model, it was believed that parents who both had blue eyes could not have a child with brown eyes. However, later studies showed that this model was too simplistic. Although it is uncommon, parents with blue eyes can have children with brown eyes. The inheritance of eye color is more complex than originally suspected because multiple genes are involved. While a child's eye color can often be predicted by the eye colors of his or her parents and other relatives, genetic variations sometimes produce unexpected results.

Several disorders that affect eye color have been described. Ocular albinism is characterized by severely reduced pigmentation of the iris, which causes very light-colored eyes and significant problems with vision. Another condition called oculocutaneous albinism affects the pigmentation of the skin and hair in addition to the eyes. Affected individuals tend to have very light-colored irises, fair skin, and white or light-colored hair. Both ocular albinism and oculocutaneous albinism result from mutations in genes involved in the production and storage of melanin. Another condition called heterochromia is characterized by different-colored eyes in the same individual. Heterochromia can be caused by genetic changes or by a problem during eye development, or it can be acquired as a result of a disease or injury to the eye.

Scientific journal articles for further reading

Sturm RA, Duffy DL, Zhao ZZ, Leite FP, Stark MS, Hayward NK, Martin NG, Montgomery GW. A single SNP in an evolutionary conserved region within intron 86 of the *HERC2* gene determines human blue-brown eye color. Am J Hum Genet. 2008 Feb;82(2):424-31. doi: 10.1016/j.ajhg.2007.11.005. Epub 2008 Jan 24. PubMed: 18252222 (http://www.ncbi.nlm.nih.gov/pubmed/18252222). Free full text available from PubMed Central: PMC2427173 (http://www.ncbi.nlm.nih.gov/pmc/articles/PMC2427173/).

Sturm RA, Larsson M. Genetics of human iris colour and patterns. Pigment Cell Melanoma Res. 2009 Oct;22(5):544-62. doi: 10.1111/j.1755-148X.2009.00606.x. Epub 2009 Jul 8. Review. PubMed: 19619260 (http://www.ncbi.nlm.nih.gov/pubmed/19619260).

White D, Rabago-Smith M. Genotype-phenotype associations and human eye color. J Hum Genet. 2011 Jan;56(1):5-7. doi: 10.1038/jhg.2010.126. Epub 2010 Oct 14. Review. PubMed: 20944644 (http://www.ncbi.nlm.nih.gov/pubmed/20944644)

To learn more about the genetics of eye color:

John H. McDonald at the University of Delaware discusses the myth that eye color is determined by a single gene (http://udel.edu/~mcdonald/mytheyecolor.html).

The Tech Museum of Innovation at Stanford University provides a Q&A explaining how brown-eyed parents can have blue-eyed children (http://genetics.thetech.org/ask-a-geneticist/brown-eyed-parents-blue-eyed-kids).

The University of Kansas Medical Center offers links to additional resources (https://www.kumc.edu/gec/support/eyecolor.html) about the genetics of eye and hair color.

More detailed information about ocular albinism (http://omim.org/entry/300500) and oculocutaneous albinism (http://omim.org/entry/203100), as well as the genetics of eye, hair, and skin color variation (http://omim.org/entry/227220), is available from OMIM.org.

A brief description of heterochromia (https://medlineplus.gov/ency/article/003319.htm) is available from MedlinePlus.

Is intelligence determined by genetics?

Like most aspects of human behavior and cognition, intelligence is a complex (https://ghr.nlm.nih.gov/primer/mutationsanddisorders/complexdisorders) trait that is influenced by both genetic and environmental factors.

Intelligence is challenging to study, in part because it can be defined and measured in different ways. Most definitions of intelligence include the ability to learn from experiences and adapt to changing environments. Elements of intelligence include the ability to reason, plan, solve problems, think abstractly, and understand complex ideas. Many studies rely on a measure of intelligence called the intelligence quotient (IQ).

Researchers have conducted many studies to look for genes that influence intelligence. Many of these studies have focused on similarities and differences in IQ within families, particularly looking at adopted children and twins. These studies suggest that genetic factors underlie about 50 percent of the difference in intelligence among individuals. Other studies have examined variations across the entire genomes of many people (an approach called genome-wide association studies (https://ghr.nlm.nih.gov/primer/genomicresearch/gwastudies) or GWAS) to determine whether any specific areas of the genome are associated with IQ. These studies have not conclusively identified any genes that underlie differences in intelligence. It is likely that a large number of genes are involved, each of which makes only a small contribution to a person's intelligence.

Intelligence is also strongly influenced by the environment. Factors related to a child's home environment and parenting, education and availability of learning resources, and nutrition, among others, all contribute to intelligence. A person's environment and genes influence each other, and it can be challenging to tease apart the effects of the environment from those of genetics. For example, if a child's IQ is similar to that of his or her parents, is that similarity due to genetic factors passed down from parent to child, to shared environmental factors, or (most likely) to a combination of both? It is clear that both environmental and genetic factors play a part in determining intelligence.

Scientific journal articles for further reading

Deary IJ. Intelligence. Curr Biol. 2013 Aug 19;23(16):R673-6. doi: 10.1016/j.cub.2013.07.021. PubMed: 23968918 (http://www.ncbi.nlm.nih.gov/pubmed/23968918). Free full-text available from the publisher: http://www.sciencedirect.com/science/article/pii/S0960982213008440

Deary IJ, Johnson W, Houlihan LM. Genetic foundations of human intelligence. Hum Genet. 2009 Jul;126(1):215-32. doi: 10.1007/s00439-009-0655-4. Epub

2009 Mar 18. Review. PubMed: 19294424 (http://www.ncbi.nlm.nih.gov/
pubmed/19294424).

Plomin R, Deary IJ. Genetics and intelligence differences: five special findings.
Mol Psychiatry. 2015 Feb;20(1):98-108. doi: 10.1038/mp.2014.105. Epub
2014 Sep 16. Review. PubMed: 25224258 (http://www.ncbi.nlm.nih.gov/
pubmed/25224258). Free full-text available from PubMed Central: PMC4270739
(http://www.ncbi.nlm.nih.gov/pmc/articles/PMC4270739/).

Sternberg RJ. Intelligence. Dialogues Clin Neurosci. 2012 Mar;14(1):19-27.
Review. PubMed: 22577301 (http://www.ncbi.nlm.nih.gov/pubmed/22577301).
Free full-text available from PubMed Central: PMC3341646 (http://
www.ncbi.nlm.nih.gov/pmc/articles/PMC3341646/)

To find out more about the influence of genetics on intelligence:

This news release from the journal Nature explains why it is so difficult to identify
genes associated with IQ: "'Smart genes' prove elusive" (http://www.nature.com/
news/smart-genes-prove-elusive-1.15858) (September 8, 2014)

The Tech Museum of Innovation at Stanford University provides a Q&A about
the influence of genes and environment on IQ (http://genetics.thetech.org/ask-a-
geneticist/intelligence-and-genetics).

The Cold Spring Harbor Laboratory offers an interactive tool called Genes to
Cognition (http://www.g2conline.org/) that provides information about many
aspects of the genetics of neuroscience.

Is handedness determined by genetics?

Like most aspects of human behavior, handedness is a complex (https://ghr.nlm.nih.gov/primer/mutationsanddisorders/complexdisorders) trait that appears to be influenced by multiple factors, including genetics, environment, and chance.

Handedness, or hand preference, is the tendency to be more skilled and comfortable using one hand instead of the other for tasks such as writing and throwing a ball. Although the percentage varies by culture, in Western countries 85 to 90 percent of people are right-handed and 10 to 15 percent of people are left-handed. Mixed-handedness (preferring different hands for different tasks) and ambidextrousness (the ability to perform tasks equally well with either hand) are uncommon.

Hand preference begins to develop before birth. It becomes increasingly apparent in early childhood and tends to be consistent throughout life. However, little is known about its biological basis. Hand preference probably arises as part of the developmental process that differentiates the right and left sides of the body (called right-left asymmetry). More specifically, handedness appears to be related to differences between the right and left halves (hemispheres) of the brain. The right hemisphere controls movement on the left side of the body, while the left hemisphere controls movement on the right side of the body.

It was initially thought that a single gene controlled handedness. However, more recent studies suggest that multiple genes, perhaps up to 40, contribute to this trait. Each of these genes likely has a weak effect by itself, but together they play a significant role in establishing hand preference. Studies suggest that at least some of these genes help determine the overall right-left asymmetry of the body starting in the earliest stages of development.

So far, researchers have identified only a few of the many genes thought to influence handedness. For example, the PCSK6 gene has been associated with an increased likelihood of being right-handed in people with the psychiatric disorder schizophrenia. Another gene, LRRTM1, has been associated with an increased chance of being left-handed in people with dyslexia (a condition that causes difficulty with reading and spelling). It is unclear whether either of these genes is related to handedness in people without these conditions.

Studies suggest that other factors also contribute to handedness. The prenatal environment and cultural influences may play a role. Additionally, a person's hand preference may be due partly to random variation among individuals.

Like many complex traits, handedness does not have a simple pattern of inheritance. Children of left-handed parents are more likely to be left-handed than are children of right-handed parents. However, because the overall chance of being left-handed is relatively low, most children of left-handed parents are right-handed. Identical twins are more likely than non-identical twins (or other siblings) to be either right-handed or left-handed, but many twins have opposite hand preferences.

Scientific journal articles for further reading

Armour JA, Davison A, McManus IC. Genome-wide association study of handedness excludes simple genetic models. Heredity (Edinb). 2014 Mar;112(3):221-5. doi:10.1038/hdy.2013.93. Epub 2013 Sep 25. PubMed: 24065183 (http://www.ncbi.nlm.nih.gov/pubmed/24065183). Free full-text available from PubMed Central: PMC3931166 (http://www.ncbi.nlm.nih.gov/pmc/articles/PMC3931166/).

Brandler WM, Morris AP, Evans DM, Scerri TS, Kemp JP, Timpson NJ, St Pourcain B, Smith GD, Ring SM, Stein J, Monaco AP, Talcott JB, Fisher SE, Webber C, Paracchini S. Common variants in left/right asymmetry genes and pathways are associated with relative hand skill. PLoS Genet. 2013;9(9):e1003751. doi: 10.1371/journal.pgen.1003751. Epub 2013 Sep 12. PubMed: 24068947 (http://www.ncbi.nlm.nih.gov/pubmed/24068947). Free full-text available from PubMed Central: PMC3772043 (http://www.ncbi.nlm.nih.gov/pmc/articles/PMC3772043/).

Brandler WM, Paracchini S. The genetic relationship between handedness and neurodevelopmental disorders. Trends Mol Med. 2014 Feb;20(2):83-90. doi: 10.1016/j.molmed.2013.10.008. Epub 2013 Nov 23. Review. PubMed: 24275328 (http://www.ncbi.nlm.nih.gov/pubmed/24275328). Free full-text available from PubMed Central: PMC3969300 (http://www.ncbi.nlm.nih.gov/pmc/articles/PMC3969300/).

McManus IC, Davison A, Armour JA. Multilocus genetic models of handedness closely resemble single-locus models in explaining family data and are compatible with genome-wide association studies. Ann N Y Acad Sci. 2013 Jun;1288:48-58. doi:10.1111/nyas.12102. Epub 2013 Apr 30. PubMed: 23631511 (http://www.ncbi.nlm.nih.gov/pubmed/23631511). Free full-text available from PubMed Central: PMC4298034 (http://www.ncbi.nlm.nih.gov/pmc/articles/PMC4298034/).

To find out more about how handedness is determined:

General information about left-handedness, including its causes, is available from the Better Health Channel (https://www.betterhealth.vic.gov.au/health/healthyliving/left-handedness) (Australia).

The Genetic Literacy Project provides a discussion of genetic factors related to handedness (http://www.geneticliteracyproject.org/2014/12/02/left-handedness-genes-and-a-matter-of-chance/).

The Washington State Twin Registry has an FAQ about hand preference in identical twins (http://wstwinregistry.org/2015/10/01/do-identical-twins-always-have-the-same-hand-preference/).

The Max Planck Institute for Psycholinguistics in the Netherlands is carrying out a study of genetics and handedness in the general population (http://www.mpi.nl/departments/language-and-genetics/projects/brain-and-behavioural-asymmetries/genetics-of-handedness).

Is the probability of having twins determined by genetics?

The likelihood of conceiving twins is a complex (https://ghr.nlm.nih.gov/primer/ mutationsanddisorders/complexdisorders) trait. It is probably affected by multiple genetic and environmental factors, depending on the type of twins. The two types of twins are classified as monozygotic and dizygotic.

Monozygotic (MZ) twins, also called identical twins, occur when a single egg cell is fertilized by a single sperm cell. The resulting zygote splits into two very early in development, leading to the formation of two separate embryos. MZ twins occur in 3 to 4 per 1,000 births worldwide. Research suggests that most cases of MZ twinning are not caused by genetic factors. However, a few families with a larger-than-expected number of MZ twins have been reported, which indicates that genetics may play a role. It is possible that genes involved in sticking cells together (cell adhesion) may contribute to MZ twinning, although this hypothesis has not been confirmed. Most of the time, the cause of MZ twinning is unknown.

Dizygotic (DZ) twins, also called fraternal twins, occur when two egg cells are each fertilized by a different sperm cell in the same menstrual cycle. DZ twins are about twice as common as MZ twins, and they are much more likely to run in families. Compared with the general population, women with a mother or sister who have had DZ twins are about twice as likely to have DZ twins themselves.

DZ twinning is thought to be a result of hyperovulation, which is the release of more than one egg in a single menstrual cycle. To explain how DZ twinning can run in families, researchers have looked for genetic factors that increase the chance of hyperovulation. However, studies examining the contributions of specific genes have had mixed and conflicting results. Few specific genes in humans have been definitively linked with hyperovulation or an increased probability of DZ twinning.

Other factors known to influence the chance of having DZ twins include the mother's age, ethnic background, diet, body composition, and number of other children. Assisted reproductive technologies such as in vitro fertilization (IVF) are also associated with an increased frequency of DZ twins.

Scientific journal articles for further reading

Hoekstra C, Zhao ZZ, Lambalk CB, Willemsen G, Martin NG, Boomsma DI, Montgomery GW. Dizygotic twinning. Hum Reprod Update. 2008 Jan-Feb;14(1):37-47. Epub 2007 Nov 16. Review. PubMed: 18024802 (http://www.ncbi.nlm.nih.gov/pubmed/18024802).

Machin G. Familial monozygotic twinning: a report of seven pedigrees. Am J Med Genet C Semin Med Genet. 2009 May 15;151C(2):152-4. doi: 10.1002/ajmg.c.30211. PubMed: 19363801 (http://www.ncbi.nlm.nih.gov/pubmed/19363801).

Mbarek H, Steinberg S, Nyholt DR, Gordon SD, Miller MB, McRae AF, Hottenga JJ, Day FR, Willemsen G, de Geus EJ, Davies GE, Martin HC, Penninx BW, Jansen R, McAloney K, Vink JM, Kaprio J, Plomin R, Spector TD, Magnusson PK, Reversade B, Harris RA, Aagaard K, Kristjansson RP, Olafsson I, Eyjolfsson GI, Sigurdardottir O, Iacono WG, Lambalk CB, Montgomery GW, McGue M, Ong KK, Perry JR, Martin NG, Stefánsson H, Stefánsson K, Boomsma DI. Identification of Common Genetic Variants Influencing Spontaneous Dizygotic Twinning and Female Fertility. Am J Hum Genet. 2016 May 5;98(5):898-908. doi: 10.1016/j.ajhg.2016.03.008. Epub 2016 Apr 28. Pubmed: 27132594 (http://www.ncbi.nlm.nih.gov/pubmed/27132594).

Painter JN, Willemsen G, Nyholt D, Hoekstra C, Duffy DL, Henders AK, Wallace L, Healey S, Cannon-Albright LA, Skolnick M, Martin NG, Boomsma DI, Montgomery GW. A genome wide linkage scan for dizygotic twinning in 525 families of mothers of dizygotic twins. Hum Reprod. 2010 Jun;25(6):1569-80. doi: 10.1093/humrep/deq084. Epub 2010 Apr 8. PubMed: 20378614 (http://www.ncbi.nlm.nih.gov/pubmed/20378614). Free full-text available from PubMed Central: PMC2912534 (http://www.ncbi.nlm.nih.gov/pmc/articles/PMC2912534/).

Shur N. The genetics of twinning: from splitting eggs to breaking paradigms. Am J Med Genet C Semin Med Genet. 2009 May 15;151C(2):105-9. doi: 10.1002/ajmg.c.30204. PubMed: 19363800 (http://www.ncbi.nlm.nih.gov/pubmed/19363800).

To learn more about the genetics of twinning:

Information about factors influencing MZ and DZ twinning is available from the Washington State Twin Registry:

- Twins run in my family. Do I have an increased chance of having twins? (http://wstwinregistry.org/2015/10/01/twins-run-in-my-family-do-i-have-an-increased-chance-of-having-twins/)
- I am a twin. Do I have an increased chance of having twins? (http://wstwinregistry.org/2015/10/01/i-am-a-twin-do-i-have-an-increased-chance-of-having-twins/)
- Does identical (MZ) twinning run in families? (http://wstwinregistry.org/2015/10/01/does-mz-twinning-run-in-families/)
- What factors are related to fraternal (DZ) twinning? (http://wstwinregistry.org/2015/10/01/what-factors-are-related-to-dz-twinning/)

The Tech from Stanford University offers a discussion of why twins can run in families (http://genetics.thetech.org/ask-a-geneticist/twin-genetics).

A brief overview of the factors that influence twinning (http://www.nhs.uk/chq/Pages/2550.aspx?CategoryID=54) is available from the UK National Health Service.

The Netherlands Twin Register provides an overview of international research on the genetics of DZ and MZ twinning (http://www.tweelingenregister.org/en/research/current-research/searching-for-twinning-genes/).

More detailed information about genetic factors related to MZ twinning (http://www.omim.org/entry/276410) and DZ twinning (http://www.omim.org/entry/276400) is available from OMIM.org.

The International Society for Twin Studies provides a list of twin registries worldwide (http://www.twinstudies.org/information/twinregisters/) and other organizations for twins and their families (http://www.twinstudies.org/information/worldwide-organizations/).

Is hair texture determined by genetics?

Genetic factors appear to play a major role in determining hair texture—straight, wavy, or curly—and the thickness of individual strands of hair. Studies suggest that different genes influence hair texture and thickness in people of different ethnic backgrounds. For example, normal variations (polymorphisms) in two genes, *EDAR* and *FGFR2*, have been associated with differences in hair thickness in Asian populations. A polymorphism in another gene, TCHH, appears to be related to differences in hair texture in people of northern European ancestry. It is likely that many additional genes contribute to hair texture and thickness in various populations.

Several genetic syndromes are characterized by unusual hair texture. These syndromes are caused by mutations in genes that play roles in hair structure and stability, including genes associated with desmosomes (specialized cell structures that hold hair cells together), keratins (proteins that provide strength and resilience to hair strands), and chemical signaling pathways involving a molecule called lysophosphatidic acid (LPA), which promotes hair growth. Genetic syndromes that feature altered hair texture include:

- Autosomal recessive hypotrichosis (caused by mutations in the *DSG4*, *LIPH*, or *LPAR6* gene)
- Keratoderma with woolly hair (caused by mutations in the *JUP*, *DSP*, *DSC2*, or *KANK2* gene)
- Monilethrix (caused by mutations in the *DSG4*, *KRT81*, *KRT83*, or *KRT86* gene)

Researchers speculate that the genes associated with these disorders probably also contribute to normal variations in hair texture and thickness, although little is known about the roles these genes play in normal hair.

Factors other than genetics can also influence hair texture and thickness. Hormones, certain medications, and chemicals such as hair relaxers can alter the characteristics of a person's hair. Hair texture and thickness can also change with age.

Scientific journal articles for further reading

Fujimoto A, Kimura R, Ohashi J, Omi K, Yuliwulandari R, Batubara L, Mustofa MS, Samakkarn U, Settheetham-Ishida W, Ishida T, Morishita Y, Furusawa T, Nakazawa M, Ohtsuka R, Tokunaga K. A scan for genetic determinants of human hair morphology: *EDAR* is associated with Asian hair thickness. Hum Mol Genet. 2008 Mar 15;17(6):835-43. Epub 2007 Dec 8. PubMed: 18065779 (http://www.ncbi.nlm.nih.gov/pubmed/18065779).

Fujimoto A, Nishida N, Kimura R, Miyagawa T, Yuliwulandari R, Batubara L, Mustofa MS, Samakkarn U, Settheetham-Ishida W, Ishida T, Morishita Y, Tsunoda T, Tokunaga K, Ohashi J. *FGFR2* is associated with hair thickness in Asian populations. J Hum Genet. 2009 Aug;54(8):461-5. doi: 10.1038/jhg.2009.61. Epub 2009 Jul 10. PubMed: 19590514 (http://www.ncbi.nlm.nih.gov/pubmed/19590514).

Medland SE, Nyholt DR, Painter JN, McEvoy BP, McRae AF, Zhu G, Gordon SD, Ferreira MA, Wright MJ, Henders AK, Campbell MJ, Duffy DL, Hansell NK, Macgregor S, Slutske WS, Heath AC, Montgomery GW, Martin NG. Common variants in the trichohyalin gene are associated with straight hair in Europeans. Am J Hum Genet. 2009 Nov;85(5):750-5. doi: 10.1016/j.ajhg.2009.10.009. Epub 2009 Nov 5. PubMed: 19896111 (http://www.ncbi.nlm.nih.gov/pubmed/19896111); PubMed Central: PMC2775823 (http://www.ncbi.nlm.nih.gov/pmc/articles/PMC2775823/).

Shimomura Y, Christiano AM. Biology and genetics of hair. Annu Rev Genomics Hum Genet. 2010;11:109-32. doi: 10.1146/annurev-genom-021610-131501. Review. PubMed: 20590427 (http://www.ncbi.nlm.nih.gov/pubmed/20590427).

To find out more about the influence of genetics on hair texture:

The Tech Museum of Innovation at Stanford University provides a Q&A on the differences in hair texture among ethnic groups (http://genetics.thetech.org/ask/ask107) and another on the inheritance of hair texture (http://genetics.thetech.org/ask/ask368).

More detailed information about the genetics of hair thickness (http://omim.org/entry/612630) and hair texture (http://omim.org/entry/139450) is available from OMIM.org.

Lister Hill National Center for Biomedical Communications
U.S. National Library of Medicine
National Institutes of Health
Department of Health & Human Services

Published on November 22, 2016

www.ingramcontent.com/pod-product-compliance
Lightning Source LLC
Chambersburg PA
CBHW081823170526
45167CB00008B/3527